# Savanna
## Safety Tips for Kids

Story by:
Lemi-Ola Erinkitola

Illustrated by:
Yuffie Yuliana

Copyright © 2020 The Critical Thinking Child LLC
All rights reserved

www.thecriticalthinkingchild.com/cybersafety
ISBN: 978-0-9899933-7-1

The third character is **Smartphone Savannah**, who goes by **Savi,** and takes the shape of a small, exciting smartphone.

She comes to Cyber City to teach kids how to use their smartphones responsibly. This is an essential skill for young learners, who may find themselves using smartphones at home, in school, or even in unsupervised situations as they get older.

The first character introduced in the Cyber City book series is **Digital Dan**, the leader od the city, who takes the form of a cloud.

He represents the cloud storage system and teaches kids how the cloud works and how to stay safe online through engaging rhymes and pictures.

The second character introduced in the Cyber City book series is **Padlock Picasso,** a talking padlock who arrives in Cyber City with a suitcase and very special paintbrush.

She immediately excites the children by showing them her scroll of password rules that she's written with her paintbrush.

The fourth character introduced in the Cyber City book series is **Appi Alex**, who takes the form of a phone app.

He is best friends with Smartphone Savi. His job is to teach kids how to choose the best apps to download.

He also shows them how to use apps safely, so they can have fun while they learn (and their parents don't have to worry).

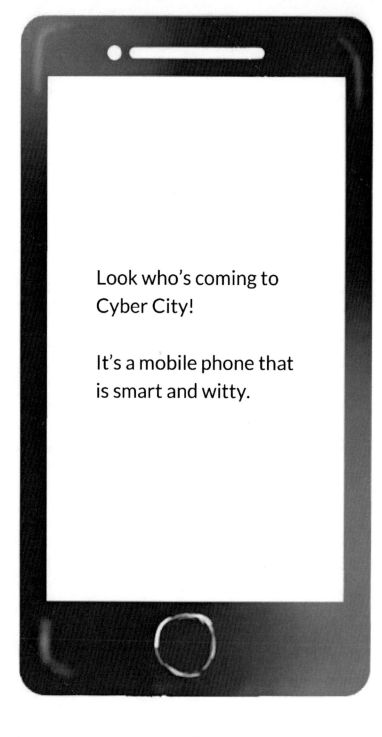

Look who's coming to Cyber City!

It's a mobile phone that is smart and witty.

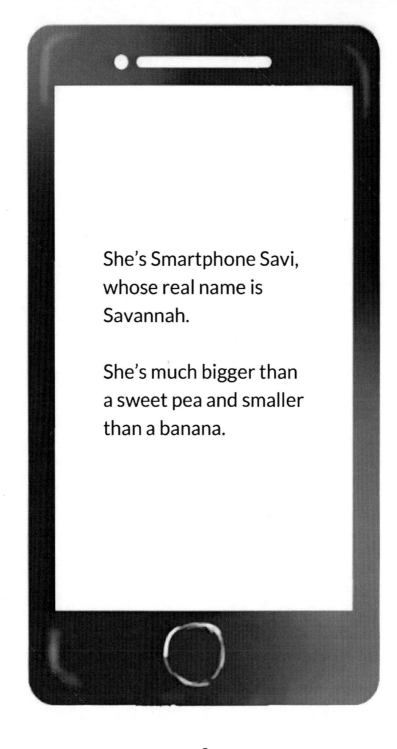

She's Smartphone Savi, whose real name is Savannah.

She's much bigger than a sweet pea and smaller than a banana.

Smartphone Savi is a very special phone, just a little bit different than the one you own.

She walks and talks and wonderfully sings,
and she'd like for kids
to know a few things.

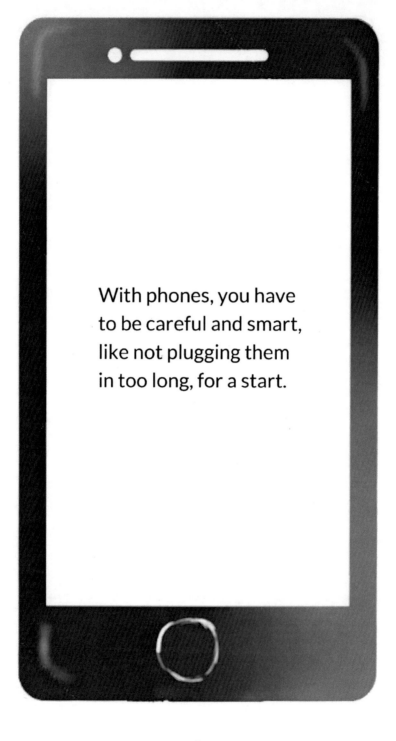

With phones, you have to be careful and smart, like not plugging them in too long, for a start.

Use your phone to call family far away, or play games and learn new things every day.

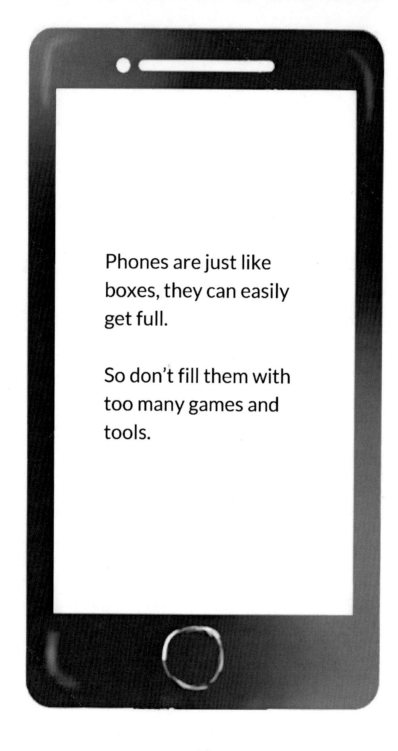

Phones are just like boxes, they can easily get full.

So don't fill them with too many games and tools.

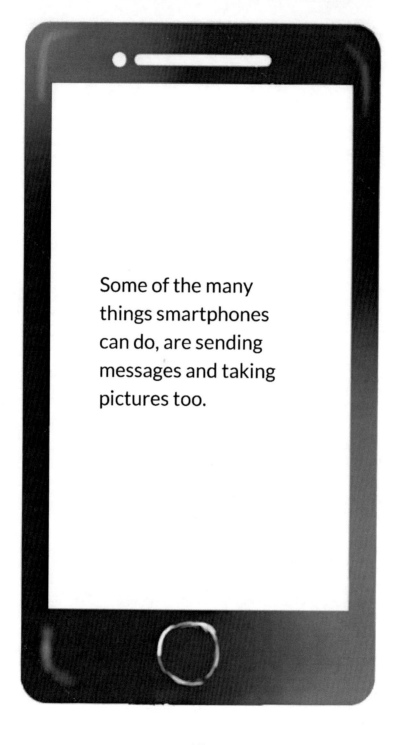

Some of the many things smartphones can do, are sending messages and taking pictures too.

You can show and share them with a friend,
but you have to be very careful of what you send.

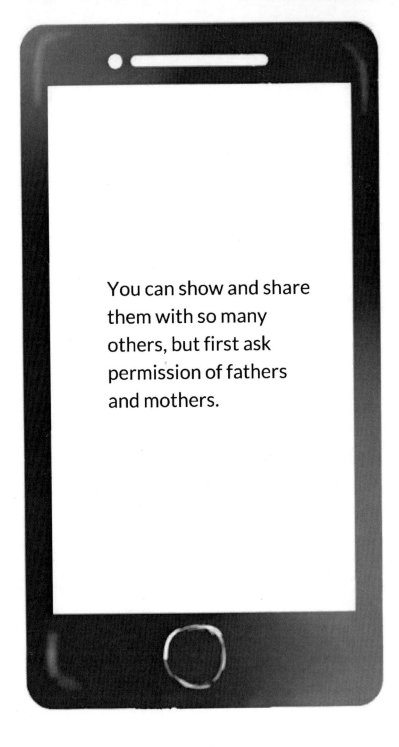

You can show and share them with so many others, but first ask permission of fathers and mothers.

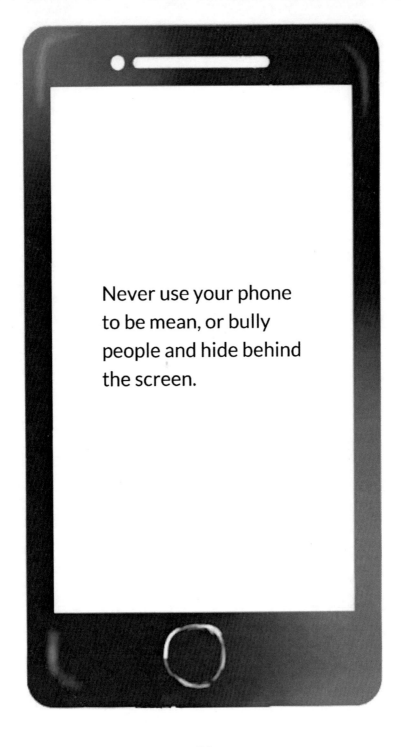

Never use your phone to be mean, or bully people and hide behind the screen.

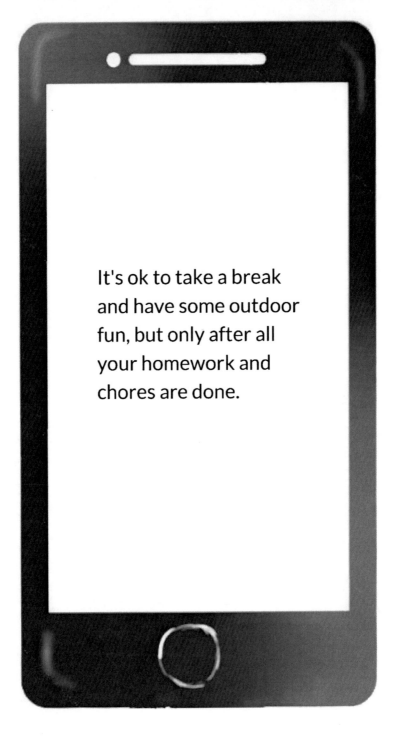

It's ok to take a break and have some outdoor fun, but only after all your homework and chores are done.

Savi just wanted to tell the little girls and boys, that smartphones are not really like toys.

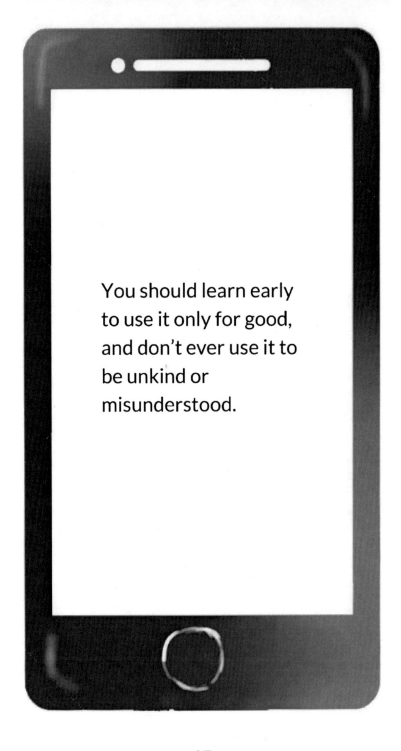

You should learn early to use it only for good, and don't ever use it to be unkind or misunderstood.

Every now and then, you should take a break, play a sport or exercise for goodness sake.

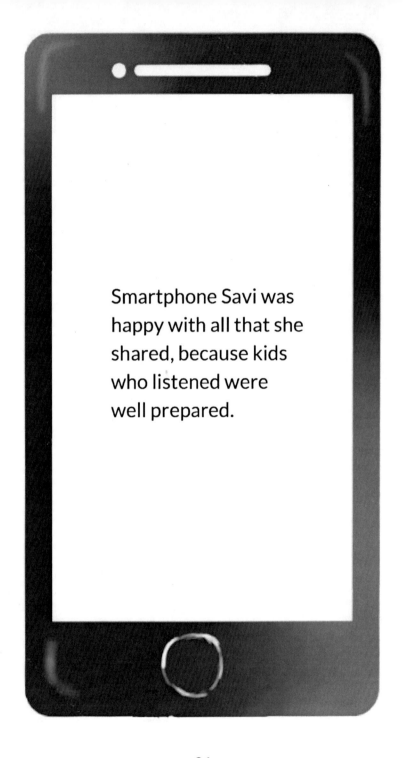

Smartphone Savi was happy with all that she shared, because kids who listened were well prepared.

**Words to Know:**

| | |
|---|---|
| bully | a person who seeks to harm for frighten someone |
| cyber | a digital world of the online world of computer networks |
| Cyber City | a place where Smartphone Savi and friends live and work |
| cybersecurity | things done online to make people and places safe |
| digital | something that uses computer technology |
| information | what you know about something or someone |
| internet | something that connects computers throughout the world |
| message | communication sent to or left for someone who cannot be contacted immediately |
| permission | give agreement for something to happen |
| phone | a device used to send voices, information or pictures over a distance |
| plugging | to connect |
| storage | space for keeping something for future use. |
| tool | a thing used to carry out an activity or job |
| password | a secret numbers, letters or symbols |
| protect | to block from harm, risk or danger |
| witty | to be funny and smart |

Made in United States
North Haven, CT
15 August 2022